Noise that Enters your Home

Understanding What it Takes to Fix Intrusive Noise

by

Graham Custard

Contents

Why this book?

The chances are you are reading this book either because you have a noise problem at home that you want to solve, or you want to deal with a noise problem on behalf of someone else. This is helpful, because I wrote this book for you.

I want to help you to understand how to deal with unwanted noise in your home and what it may involve. I have been dealing with the subject for a long time (over 30 years!) and I am concerned that many people's expectations are either unrealistic, or they are being misled by 'experts' or by trades people who have less knowledge of the subject than they might claim. There is a lot of information available out there, especially on the internet and on YouTube. Some is very good, but some is misleading.

I believe that if I can provide you with background knowledge about the subject, you will be better equipped to deal with your problem, to negotiate with traders, to spot the devious answers, to appreciate honest appraisals – and you won't have unrealistic expectations of a solution. There is no such thing as 'sound proofing,' nor perfect silence in the practical world in which we live.

If you invite a builder to undertake remedial work for you, or if you wish to do it yourself, it is an advantage to have a better understanding of what is likely to be involved. A sound knowledge (yes, it's a pun) of this subject will likely save you money or, at the very least, will save you wasting money.

Do not worry if the subject matter seems technical. The first three chapters are designed to give you an overall understanding of the subject very quickly.

Then you can dip into the chapters that are particularly relevant to your circumstances when you need them.

The Tables in Chapter 11 provide a summary of the key details discussed from Chapter 4 onwards.

Graham Custard, B.Sc., M.Sc.

1

Describing Noise

Sound and Noise

The terms 'sound' and 'noise' are used interchangeably to describe the same process; we hear it as the vibration of air. Usually, 'noise' means an unwanted sound.

A sound is heard because our ears detect the vibration of the air around us. The source of the sound disturbs the air by vibrating it and that disturbance disperses at about 770 miles per hour (or 340 meters per second). The air itself does not move at this speed, but the vibration of the air molecules is passed through the air at this rate, molecule to molecule, at what is called the speed of sound.

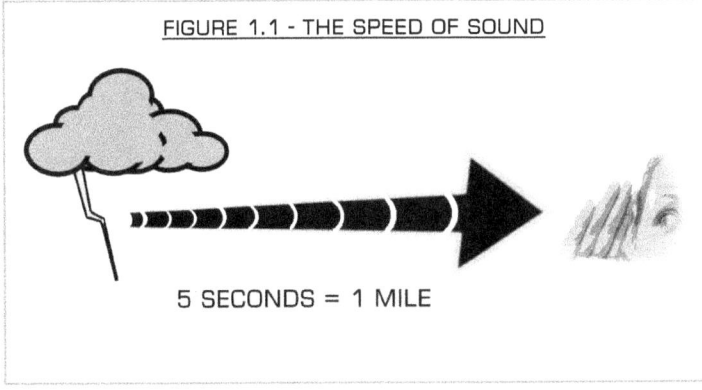

FIGURE 1.1 - THE SPEED OF SOUND

5 SECONDS = 1 MILE

It travels about a mile in 5 seconds, or a kilometer in 3 seconds (Figure 1.1). As soon as you see a flash of lightning, count the number of seconds until you hear the thunder and divide by 5; that is the distance from the flash in miles. (If you use kilometers, divide the time between the flash and the sound by 3.)

Fortunately, we are not concerned with how fast sound travels when we are dealing with the noise it generates inside our homes, but we do need to know two qualities about the air vibration that causes the noise: the intensity of the vibration (by how much the air molecules are disturbed) and its frequency, or rate of vibration.

Decibels and Frequency

The intensity of the sound is measured in decibels (dB), the frequency is measured in Hertz (Hz).

The human ear has a wide range of ability to detect the air vibrations; it can detect the faint chirrups of an insect at a great distance, and sound as intense as a gunshot (though not necessarily without pain or even damage). This represents an intensity range which is measured in multiple billions. Faced with such an enormous range of units, scientists and engineers use decibels (dB) to describe sound levels in convenient numbers. For normal human hearing, the scale starts at 0 dB and proceeds upwards to the loudest sounds we can tolerate without immediate damage at about 140 dB.

There is also the need to describe the frequency of the sound, which is measured in Hertz (Hz); a Hertz is one complete vibration of the air every second.

4

Healthy ears can hear sounds from as low as 20 Hz (20 vibrations per second) to as high as 20 kHz (20,000 vibrations per second), although our ability to hear those higher pitched sounds decreases greatly as we get older. Our ears are most sensitive to the middle range of frequencies used for speech and communication, rather than very high-pitched sounds (we cannot hear bats) and extremely low frequency sounds (deep rumbles that we feel, rather than hear). Figure 1.2 shows the range of human hearing in terms of frequency, measured in Hz.

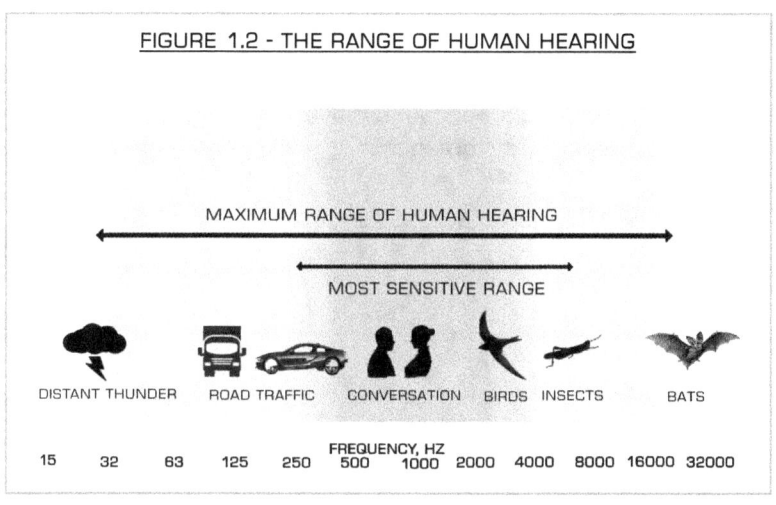

The sounds that we hear are usually very complex, because they are composed of air vibrations at multiple frequencies. Everyday sounds consist of a combination of low vibration frequencies (bass sounds in musical terms), middle frequencies and high frequencies. The air is perpetually disturbed by these vibrations at all these frequencies.

Our ears (and inner ears) detect these multiple vibrations, determine their intensity and which frequencies are present, then transmit the information to our brains for further analysis and interaction with our memory. In this way, even with our eyes closed, we can assess how

5

loud a sound appears to be and can deduce if we are listening to a child crying, a person singing, a car passing on the road, a bird, a plane, and so on.

You will often come across more than plain decibels. The intensity and frequency content of measured sound can be combined into a single number, called A-weighted decibels, expressed as dBA.

Measurements in terms of dBA give emphasis to the middle range of hearing to exclude much of the sound at very low and very high frequencies which we do not hear so well. The effects of noise on people, whether for annoyance or damage to hearing, are expressed in terms of dBA (Figure 1.3).

In our homes, we commonly encounter levels of sound between about 35 and 75 dBA, sometimes more than this.

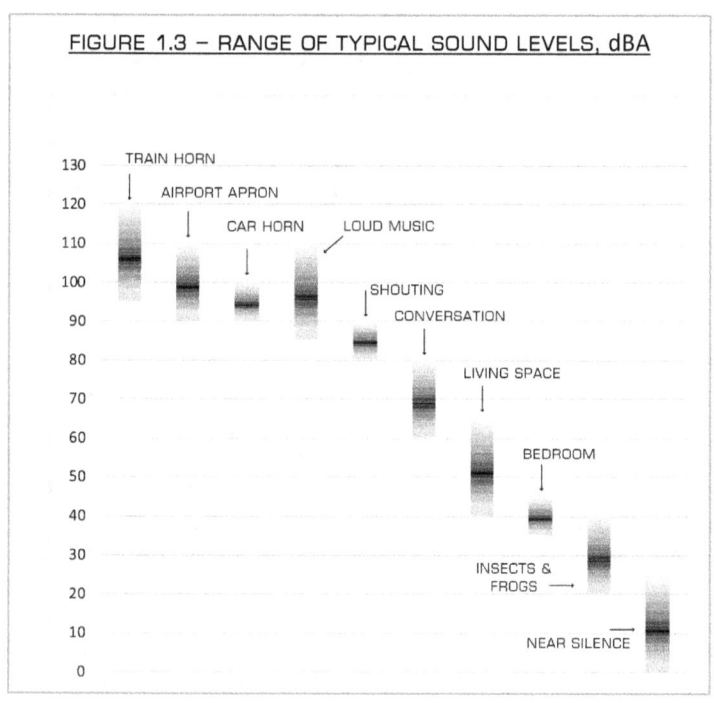

FIGURE 1.3 – RANGE OF TYPICAL SOUND LEVELS, dBA

The decibel scale is a multiplying scale; it is condensing a wide range of intensities from the millions into a little more than about a hundred decibels.

Each time the scale increases by 10 dB, the intensity of the air vibration increases 10 times. The principle is the same, whether the sound is measured in terms of dBA or just plain decibels, dB. So, a sound measured to be 60 dBA has 10 times more intensity than sound at 50 dBA, a hundred times more than sound at 40 dBA, a thousand times more than a sound at 30 dBA, and so on.

The combination of decibel levels is confusing when you are not used to handling the numbers. When the vibration intensity of a sound is doubled, it increases by only 3 dB on the decibel scale. Similarly, a halving of its intensity reduces it by 3 dB. Therefore, two equal sounds at 50 dB, would combine to produce a total of 53 dB, not 100 dB. Ten equal sources, each producing 50 dB, would raise the total level to 60 dB.

However, even though an increase in sound level of 3 dB represents a doubling of sound intensity, it is not noticeable. A change in sound level of 3 dB is not noticed by most people. We perceive its 'loudness' as not changing. Table 1.1 (next page) summarizes how we perceive changes in decibel levels.

TABLE 1.1 - HOW NOTICEABLE IS A CHANGE OF SOUND LEVEL?

Change in sound level, dB	Loudness Effect
3	Not Noticeable
4	Just Noticeable
5 to 6	Noticeable
7 to 9	Very Noticeable
10	Loudness Doubles (+) or Halves (-)

"A change in sound level of 3 dB is not noticed by most people."

Loudness

We use the term 'loudness' to describe our perception of sound. It is subjective because we usually compare one sound with another when describing loudness. An increase of 10 dB sounds twice as loud as before, a reduction of 10 dB sounds half as loud as before.

That presents challenges when attempts are made to reduce noise, because each time it is reduced by 10 dB, it only seems like half the job has been achieved. The noise seems only half as loud as before, even though the intensity of the air vibration has reduced by a factor of 10. Similarly, decreasing by 20 dB is perceived as only a quarter reduction in loudness, even though the measured intensity has dropped by a factor of 100.

"An increase of 10 dB sounds twice as loud as before,

A reduction of 10 dB sounds half as loud as before."

Companies selling you noise control products or services sometimes use this confusion about loudness to convince you their product is fantastic. Consider these statements used in advertisements:

- *"Cuts the noise by 90%"* - It is equivalent to a reduction of 10 dB, but the loudness has only been halved.

- *"Cuts the noise by 99%"* - This means a reduction of 20 dB, for which the loudness has been reduced only by a quarter.

- *"The noise is reduced by 95%"* - This is commonly encountered and sounds like a high number, but it only represents a reduction of 13 dB. That is only a slight, barely noticeable improvement on 10 dB, which means the noise will seem half as loud.

That is not to say that 20 dB, 13 dB, or even 10 dB reduction are poor. On the contrary, they can be useful reductions when the noise is an annoyance, but this perception of loudness explains why people are often disappointed with their results because they can *"still hear the offending noise."*

Loud speech in an adjacent room can still be heard clearly if the wall reduces the sound by 30 dB. In seeking privacy, we need to go further, aiming for at least 40 to 45 dB reduction, which means the loudness is reduced to about a twentieth of the original.

A 'comfort zone' begins when the sound is reduced by about 45 dB, when most people-generated noise cannot be heard. It may be necessary to go further when complete privacy from loud conversation and musical instruments are required, for which a reduction of 50 to 60 dB may be necessary (Table 1.2).

But nothing is 'sound proof.' If you listen very carefully, you can usually still hear the sound.

TABLE 1.2 – REQUIRED REDUCTION IN NOISE

Required reduction dB	What can be heard
30	Normal conversation heard but not understood
40	Loud speech heard but not understood
45	Most 'people-noise' cannot be heard
50	Speech inaudible; some loud music sounds heard
60	Very loud music sounds no longer audible

2

Reducing Noise

Noise can be reduced in three ways: by reflection, by absorption and by transmission (passing it through a sound resisting material).

Reflection and Absorption

Sound is reflected extremely well by hard surfaces, such as brick, cement or glass. Water is also a good reflector of sound and the surface is a poor absorber, which explains why, when you stand near the edge of a large lake, or visit the beach, you can hear boat engines or children playing in the water at a great distance. There is no soil or vegetation to scatter and absorb the sound before it reaches you.

This brings us to the concept of sound absorption. It is the partial removal of vibration intensity at each reflection. Specialized sound absorbing materials are made porous to allow the sound to penetrate the tiny holes before it is reflected. A small amount of energy is converted to heat (which is so small it is difficult to measure) and the reflected sound is decreased in its intensity (Figure 2.1).

Many reflections occur inside a building, or inside a room in your home. The sound vibrations reflect from room surfaces like light from a mirror, but they lose a tiny amount of their energy on each reflection.

This all happens at the speed of sound, of course, so a great many reflections will occur every second. The sound energy dies away and quickly becomes inaudible. An empty room, with many reflective surfaces that do not absorb much sound, appears 'lively' with echoes when you clap your hands. It may take the reflections about a second to die away completely. The introduction of sound absorptive surfaces, however, extracts more energy from the sound vibrations at each reflection and the sound dies away faster, within a fraction of a second.

Curtains and soft furnishings help to absorb the sound, as does carpet which is spread over a wide area and intercepts many reflections from the floor.

Tiles that are sound absorptive, made specially for the purpose, are useful to place on wall surfaces or ceilings, where they reduce the sound that is otherwise bounced around many times inside the room, especially when a carpet or large rugs on the floor are inconvenient.

Thick wall hangings that resemble rugs can also be used to reduce unwanted reflected sound in a room to make it feel less 'lively.'

FIGURE 2.1 – SOUND ABSORBERS REDUCE THE ENERGY OF REFLECTED SOUND

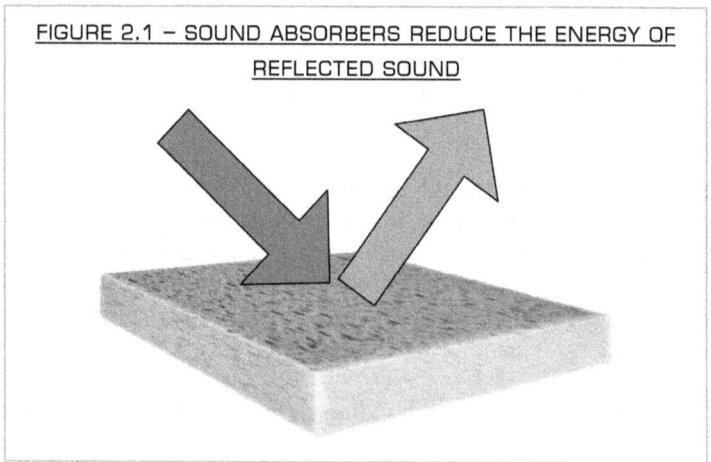

Sound absorbing materials work best at removing the higher frequency sounds. They do relatively little to stop sound passing through them. They absorb reflected sound but do not reduce transmitted sound significantly. Therefore, they are useful for controlling noise once it has entered a room, but they are not as useful for reducing noise entering the room from outside.

Sound absorption can be used to good effect inside wall cavities to remove unwanted reflections and to improve the noise reduction of the cavity, which will be discussed shortly.

Trees and shrubs absorb sound because, given sufficient distance and density of planting, their leaves and branches break up the sound with multiple reflections in all directions and thereby diminish its intensity. They become a sound absorptive material on a grand scale. Unfortunately, a thin line of trees or shrubs at home will do little to achieve any benefit.

Transmission

Dense, heavy materials, such as concrete or brick walls, reduce the sound passing through them (Figure 2.2). Increasing the weight of a wall increases its ability to resist the transmission of sound. A doubling of wall weight results in a noise reduction improvement of about 5 or 6 dB.

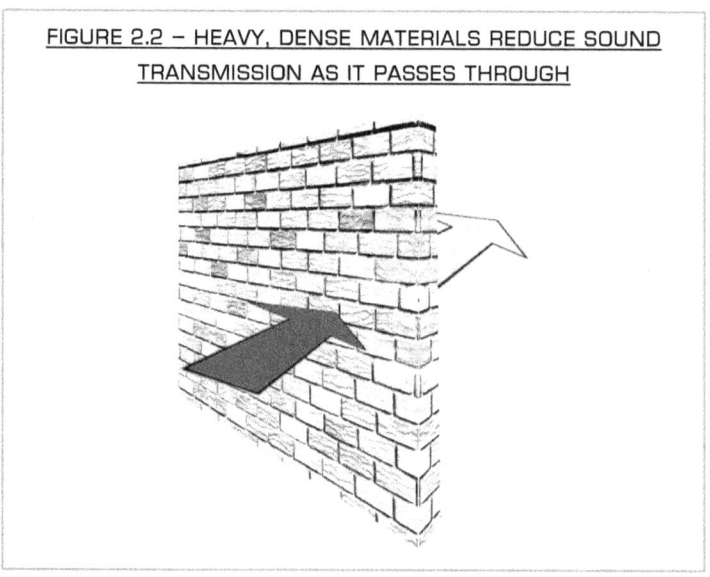

FIGURE 2.2 – HEAVY, DENSE MATERIALS REDUCE SOUND TRANSMISSION AS IT PASSES THROUGH

A glass window also reduces the sound passing through because the glass is heavy and resists the air vibration. A window with two panes of glass resists the sound better than a single pane.

Sound Transmission Class (STC)

Comparing Noise Reduction

In the USA and some other countries, suppliers of noise reducing materials use a so-called Sound Transmission Class (STC) for their products. In Europe, the Weighted Sound Reduction Index (Rw) is used instead, which is very similar to STC. The two methods rate a wall material, or a window or door, with a single number for its sound-reducing potential. That way, it is possible to compare walls, windows and doors to decide how much noise they are likely to transmit. In general, the higher the STC (or Rw) value, the better it reduces a given noise.

> *"Although a single STC number is helpful for comparing products that reduce 'people sounds,' it is misleading for dealing with low frequency sounds."*

These rating values emphasize the sound reduction at the middle and higher range of audible frequencies, the sounds that annoy us most with regards to human conversation, but they do not cover the hearing frequency range in the same way as A-weighted decibels. This is unfortunate: a wall that has a STC rating of 30, reduces the sound of human voices by approximately 30 dBA, because voices tend to be in the middle to high frequency range of our hearing, but loud music with a lot of base may only be reduced by 10 dBA.

Therefore, although a single STC (or Rw) number is helpful for comparing products that reduce 'people sounds,' it is misleading for dealing with low frequency sounds such as from heavy road traffic and commercial vehicles, washing machines on the spin cycle and loud music with a lot of bass sound. Be aware of this limitation in the values quoted.

What STC value should you aim for?

STC (and Rw) values are obtained from published laboratory measurements of wall or partition structures and cannot easily be estimated from the properties of their individual components.

For example, a drywall board may have a STC value of 20; doubling the thickness by combining two boards might reasonably suggest that the combination has a STC of 25 (because of the doubling of the weight), but test results may show otherwise. It all depends how they are put together, and on what kind of framework they are mounted.

The only way to know the STC or Rw precisely for a partition, wall or window is to have it tested.

Clearly, it is not within the resources of a home-owner or builder to submit material combinations for testing. Therefore, for the purposes of this book, I have taken a practical approach:

- Trends and values have been summarized after examining data from several published results of STC values for various wall combinations;

- The trends are used to provide practical guidelines for reducing the noise transmission into your home. (If you use Rw in your country, use these STC values as a guide.)

Just be aware that your results will vary because, unless you purchase the precise combinations of specific products that have been tested, you will not be matching those laboratory test conditions. Your results may be better, they may be worse.

A wall STC of 45 would make life comfortable between neighbors because most people-generated noise would not be heard.

It is sensible to aim for a higher figure, if possible, because the actual noise reduction achieved cannot be guaranteed and you may see lower values caused by variations of materials and fit. It is especially important to aim for higher figures if there is much low frequency noise, to achieve as much reduction of this sound as possible.

> *"A wall STC of 45 would make life comfortable between neighbors."*

In the USA, state building codes have been updated recently to require STC values of 45 for partition walls in homes. This has not always been the case and older housing stock will likely have partitions with STC values less than this.

3

Noise Enters from Everywhere

Figure 3.1, on the next page, illustrates a typical domestic room; it is smaller than a whole house, but it shows the significant routes taken by annoying noise which enters the room from outside. The same principle applies whether it is a single room, an apartment or an entire house.

It shows four main routes for external noise entering the room: through the windows, the ceiling (from a room above or through the roof), an adjoining wall (or walls) and the door. In some circumstances, noise may come through the floor from below, but this is less common as a nuisance.

Each of the arrowed sources are important, but Figure 3.2 illustrates hidden weaknesses that admit at least as much sound as doors, windows and walls. These are air gaps, which transmit the sound directly from outside to inside with very little reduction. They may seem to be very small, but their total area adds up and can become significant. Only a 5% total area of leakage around a wall or door will diminish its noise reduction to only about 15 dB, even if it were otherwise capable of 40 or 50 dB reduction!

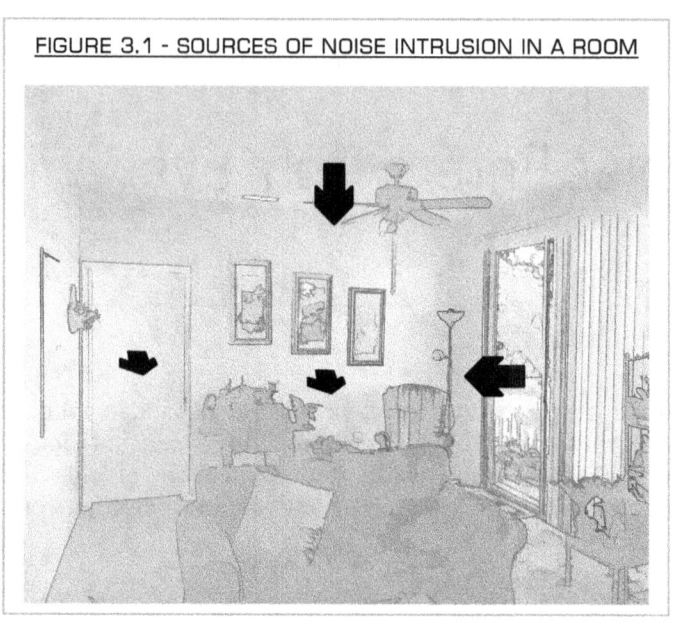

FIGURE 3.1 - SOURCES OF NOISE INTRUSION IN A ROOM

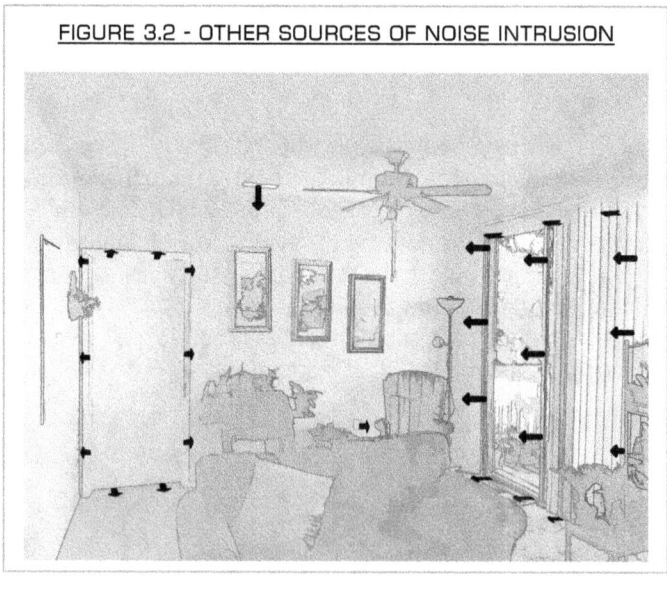

FIGURE 3.2 - OTHER SOURCES OF NOISE INTRUSION

In the illustration of Figure 3.2, there are significant air gaps which should not be left untouched for noise treatment:

- gaps around the door in its frame, and especially at the bottom where it clears the floor;

- gaps around the windows, as the frame meets the existing walls, at the closure of the windows and in any slide channels;

- ventilation openings and ducts (in this case in the ceiling) which may also provide a path for noise;

- electrical sockets which have air gaps in their structure.

4

The Walls

Walls divide living spaces from each other and are therefore the most common causes of complaint about noise intrusion. Noise from the environment outside can also cause issues after it has passed through the external walls.

The weight of a wall is important for its ability to resist sound. In general, the heavier a solid material, the better it is at reducing noise. Figure 4.1 illustrates what is known as the 'mass law.' Noise reduction

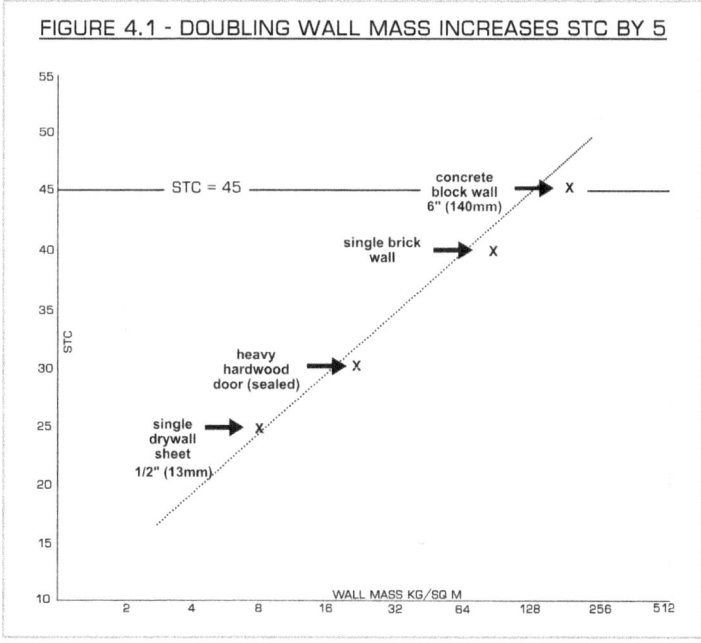

FIGURE 4.1 - DOUBLING WALL MASS INCREASES STC BY 5

improves by about 5 dB for each doubling of material weight (or mass), as shown by the diagonal line.

In that graph, only the concrete block wall meets the desirable STC=45 line. Some materials perform better than the mass law and are found above the diagonal line. On the other hand, some can be worse because they fall below the diagonal line, as shown in Figure 4.2.

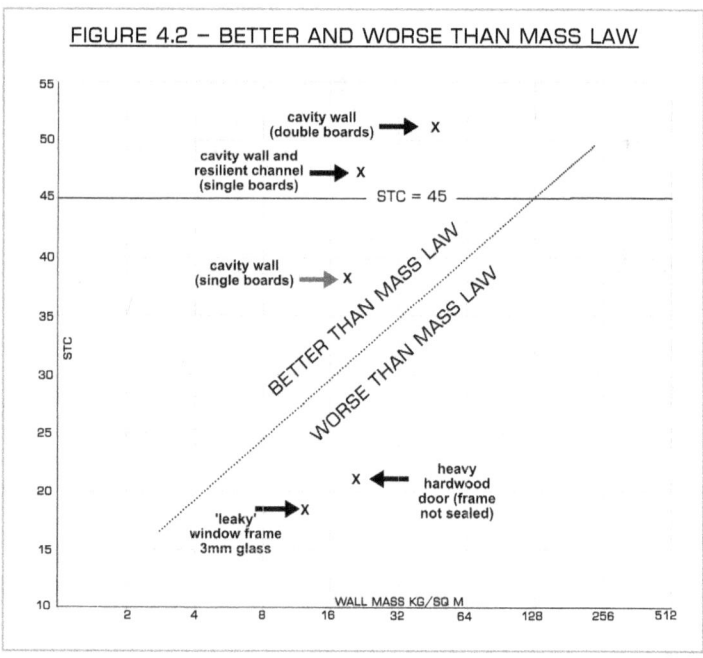

FIGURE 4.2 – BETTER AND WORSE THAN MASS LAW

A heavy wood door, which has insufficient sealing around the door frame when it is closed, is below the line because noise is leaking through gaps which are not airtight. Similarly, a poorly fitting window in its frame does not stop a lot of noise. On the other hand, when attention has been given to the frame sealing, the noise reduction improves and the STC values move closer to, or even above, the mass law line.

Cavity Walls

Cavity walls with gypsum boards (also called drywall) reduce more noise than their weight alone dictates and are found well above the mass law line. This is the reason they are so popular for noise insulating rooms in homes

Relying on weight of material alone would require thick, heavy walls to achieve high STC values, which take up space and can become expensive to build. Relatively lightweight wall boards with cavities can achieve high STC values which reach the suggested 'comfort zone' above 45, as previously shown by Figure 4.2.,

However, not all cavity walls in existing residences achieve STC values as high as 45, which often causes complaints to arise. The reasons, and how to improve them, will now be discussed. Usually, a cavity wall consists of gypsum boards mounted on both sides of a framework of wood or steel. Insulation material, such as glass fiber blanket or mineral fiber, partially fills the cavity to absorb sound reflections inside (Figure 4.3).

The basic cavity wall is a simple construction and it does not sufficiently isolate the 'people sounds' from next door because it can only achieve a STC value of about 35. Air leaks arising from poor fit of boards and frame can reduce the performance to less than 30.

Figures 4.4 (a) and (b) are overhead sketch views of this type of design, I shall use these to illustrate construction details of cavity walls, rather than draw a 3D model each time.

FIGURE 4.3 - DRYWALL PARTITION ON 2"X4" FRAME

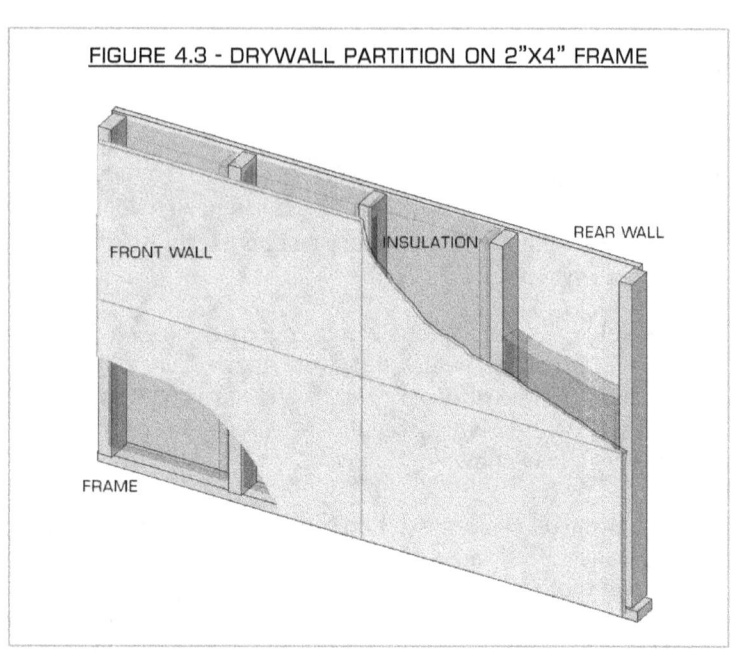

FRONT WALL

INSULATION

REAR WALL

FRAME

FIGURE 4.4 – OUTLINE SKETCH OF BASIC DRYWALL PARTITION

(A)

TIMBER FRAME 2" X 4"
(50 X 100 MM)

(B)

METAL FRAME
EQUIVALENT

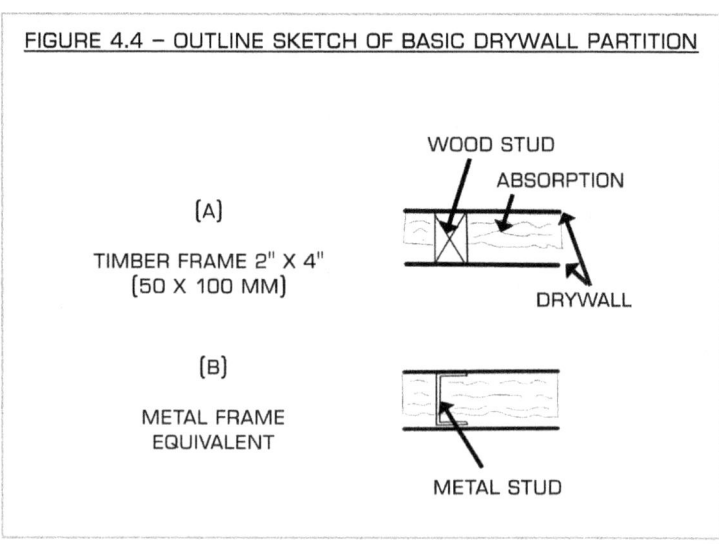

WOOD STUD

ABSORPTION

DRYWALL

METAL STUD

What can be done to improve a cavity wall?

First, understand how the sound is passing through the wall. The sound on the source side causes the gypsum panel of the wall to vibrate, the panel resists the vibration and passes a reduced sound level into the air of the cavity. The sound in turn vibrates the second panel on the receiver side which transmits a further diminished sound into the receiver room (Figure 4.5).

Unfortunately, the framework itself, in contact with both wall surfaces, transmits vibration directly from wall to wall so that the STC of the whole wall is diminished. Small air gaps at the edges of the frame readily transmit sound where it meets the adjoining wall, ceiling and floor. There may also be significant hairline gaps at the joins between wall boards, which allow sound to pass through unhindered and contribute to a diminished total STC. An electrical socket provides another opportunity for sound to leak through the wall; there may be

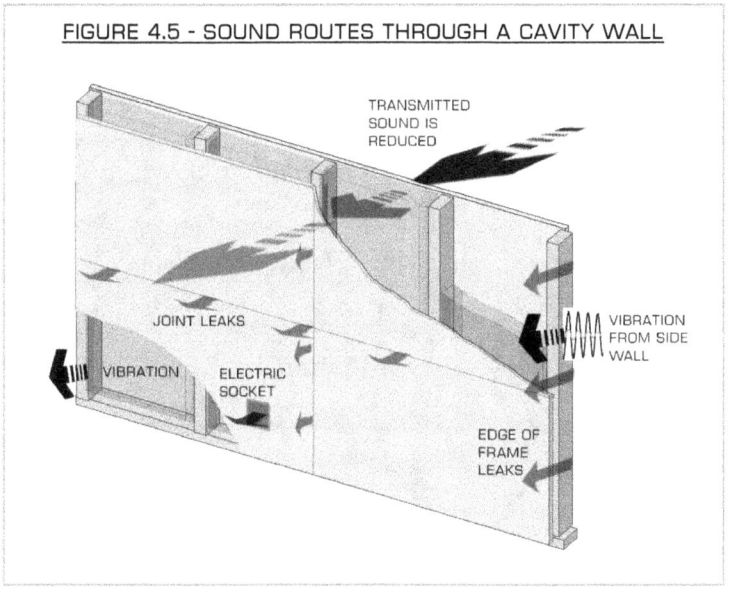

FIGURE 4.5 - SOUND ROUTES THROUGH A CAVITY WALL

two adjacent to each other, one on the source side, one in the receiving room.

If there is insufficient sound absorption in the cavity, the sound bounces around and boosts the sound level inside the cavity, which means the sound emerging on the receiver's side of the wall is that much higher than it should be.

A tempting remedy is to attach another layer of gypsum board on the receiver side. This will not do any harm, but any benefit from just increasing the weight of the wall is only small without other modifications.

For example, if the same thickness of wall panel is applied to one side, the total weight is increased only by 50%. According to the mass law, this will increase the STC by about 3 points, which is not a noticeable improvement.

You could try and improve it further by adding even more wall weight on your side, but even doubling the weight of the wall will only improve the STC by 5 points.

Two important weaknesses of the existing structure remain: air leaks and vibration. Also, the condition of the sound absorption in the cavity is unknown and may not be sufficiently thick.

It is very likely that the wall cannot achieve a STC value higher than about 30.

The voices on the other side can still be heard!

Improving the Frame

The target should be to achieve a STC of at least 45, preferably 50 if possible. The most important task is to prevent the frame from transmitting vibration from one side of the wall to the other.

> *"The most important task is to prevent the frame from transmitting vibration from one side of the wall to the other."*

Assuming the wall on the neighbor's side cannot be modified because they are living there, it is first necessary to remove the wall board on the side that is in your home, and to replace it with a wall that is less in contact with the frame. There are two ways of doing this:

- add a second framework to support the new receiver wall that does not touch the existing frame;

- attach the new wall to a slightly sprung mounting, called resilient channel, on the existing frame; the channel reduces the contact area with the frame and is supposed to reduce vibration transmission.

Both methods have advantages and disadvantages.

A Second Frame

A second wood framework is constructed, using standard 2 x 4 in. (50 x 100 mm) timber and studs with a 1 in. (25 mm) gap between the adjacent, vertical studs. It is essential that the studs do not ever contact each other, or they will transmit unwanted vibration. It has the potential to provide a STC value of approximately 53 points (Figure 4.6).

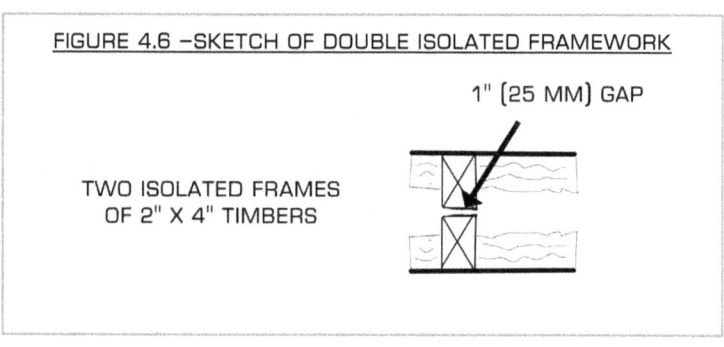

FIGURE 4.6 —SKETCH OF DOUBLE ISOLATED FRAMEWORK

1" (25 MM) GAP

TWO ISOLATED FRAMES
OF 2" X 4" TIMBERS

The STC could be boosted above 55 if a double thickness of gypsum board is added on one side.

A limitation of further noise reduction is imposed by vibration in the existing walls, floor and ceiling. It finds a path into the new framework (which must be attached to the surrounding walls and floor) and is transmitted into the receiving room as sound.

A big disadvantage of using the second framework is the sacrifice of almost 6 inches (150 mm) of space in the receiving room.

It is possible to reduce the cavity by staggering the studs and moving the frame inwards, always ensuring the studs and frames do not touch, but there is still a loss of space with this wall.

Staggered Studs

An even more space-saving alternative is to use 2 x 4 in. (100 x 100 mm) vertical studs on a 2 x 6 in. (100 x 150 mm) frame, allowing the studs for the new wall to be staggered with those of the existing wall in such a way that they never touch each other (Figures 4.7 and 4.8).

This has the potential to achieve a STC of about 45, but it can only be used in new builds or for renovation where the party wall can be knocked down and the new frame constructed.

FIGURE 4.7 – SKETCH OF STAGGERED STUDS
ON A 2" X 6" FRAME

STAGGERED 2" X4"
(50 X 100 MM)
STUDS ON 2" X 6"
(50 X 150 MM)
FRAME

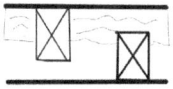

FIGURE 4.8 - STAGGERED 2"X 4" STUDS ON 2"X 6" FRAME

Stud Spacing

If a new frame is being constructed, it is possible to improve the performance slightly by using a wider stud spacing.

Test figures indicate that the STC of a cavity wall is improved by about 3 to 5 points if a wider stud spacing of 24 in. (60 cm) is used instead of the standard spacing of about 16 to 18 in. (40 to 45 cm). This is a helpful improvement if the weight of the chosen wall can be supported by this wider spacing.

Metal Studs and Frame

The frame improvements described in this chapter can also be applied to metal frames and studs. It is feasible to use a combination of metal frame for one wall and timber frame for the other, providing the structure will support the chosen weight of wall material.

Resilient Channel

An alternative to constructing a new frame is to mount the new inside wall on so-called resilient channel, which attaches to the existing frame, acts like a spring support and reduces the contact area of the new wall to the frame.

The channel, made of folded metal strip, is found in a variety of designs, of the type illustrated in Figure 4.9. The channel is screwed to the existing frame and then the wall board is screwed to the raised section of the channel, thereby separating the board from direct contact with the frame. If the channels are laid across the studs, rather than along them, there is less contact area between frame and channels, which reduces the opportunity for vibration transmission.

Test figures suggest that the use of resilient channel improves the STC of a cavity wall by 10 points compared with mounting gypsum board on the frame directly. This method therefore offers a very useful reduction in noise for cavity walls and can be used in combination with other cavity wall improvements already described.

Although the trend for reducing noise transmission through a wall with resilient channel is about 10 dB, real-world results will depend both on the type of channel used and the method of fixing, which probably differ from laboratory test conditions.

It is better to purchase channel that is of light enough gauge to be slightly flexible; a lighter gauge metal that is more 'springy' (25 gauge is usually available) is likely to yield better results than a more rigid gauge.

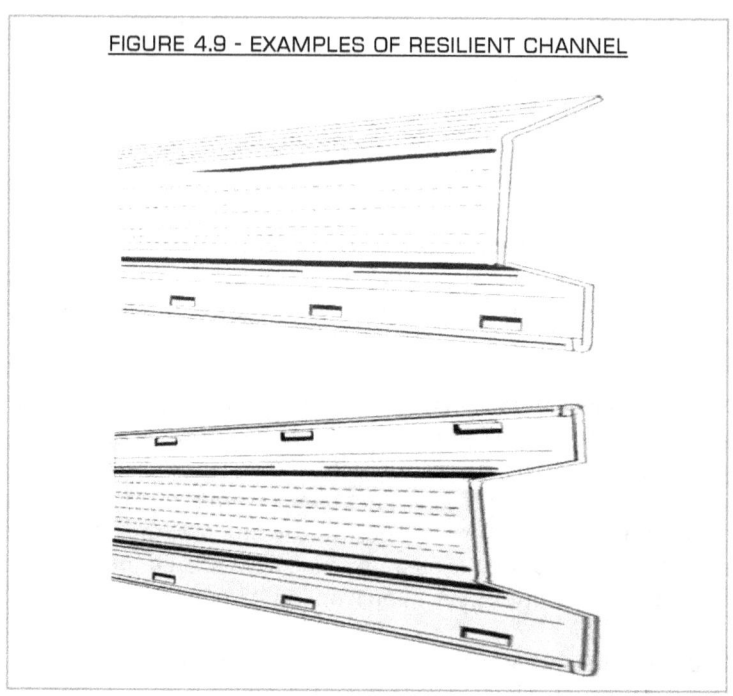

FIGURE 4.9 - EXAMPLES OF RESILIENT CHANNEL

In Figure 4.9, above, the channel is screwed horizontally onto the vertical studs of the frame through the slots. The gypsum drywall board is screwed to the raised section only.

It is essential that the screws attach the board at positions where they do not penetrate or touch the frame or wall behind, thereby short circuiting the attempts to vibration isolate the wall. It seems an obvious thing to say, but this appears to be the cause of many complaints of poor remedial work. With a little understanding, and careful supervision of the work, such pitfalls can be avoided.

34

A typical arrangement for resilient channel on an existing timber frame is shown in Figure 4.10, below. The channels are laid horizontally across the frame (rather than along the studs, vertically) to reduce direct contact of each channel with the frame as much as possible. The vertical spacing of the horizontal channels should be about 16 to 24 in. (400 to 600 mm), depending on the weight of wall to be supported. If possible, the wider spacing of 24 in. is preferred because it offers the most noise reduction.

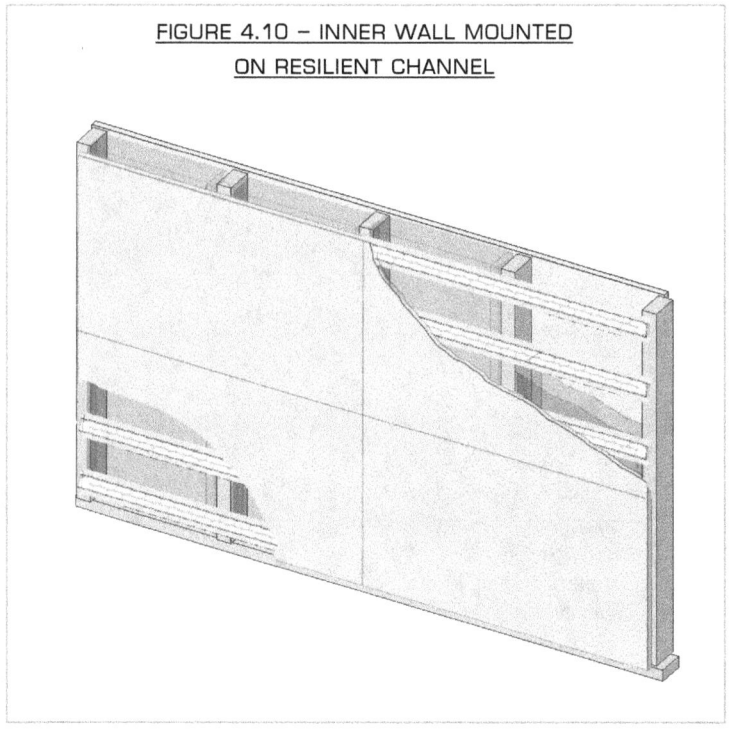

FIGURE 4.10 – INNER WALL MOUNTED
ON RESILIENT CHANNEL

5

Attention to Details

Sound Absorption in the Cavity

It is important to introduce sound absorptive material into the cavity. This soaks up reflected sound inside the cavity before it has the chance to pass through the wall. Regular glass fiber thermal insulation, or mineral fiber (e.g. Rockwool) meets the need well enough. It does not need to fill the entire cavity to its fullest depth: aim for a fill of about 75% of the cavity depth. For noise control, there is no advantage in fully packing the cavity.

Edges of the New Wall

Ensure the new wall boards (and resilient channels, if fitted) do not touch the surrounding walls, ceiling and floor. A gap of about ¼ in. (6 mm) is sufficient.

Fill the existing gaps with a flexible sealant (Figures 5.1 and 5.2). A heavy-duty sealant for external use should be applied which can maintain flexibility; some companies offer 'acoustic sealant' which is 'viscoelastic' and claims to remain flexible.

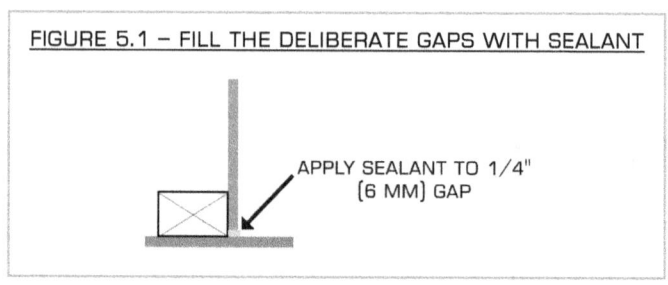

FIGURE 5.1 – FILL THE DELIBERATE GAPS WITH SEALANT

APPLY SEALANT TO 1/4"
(6 MM) GAP

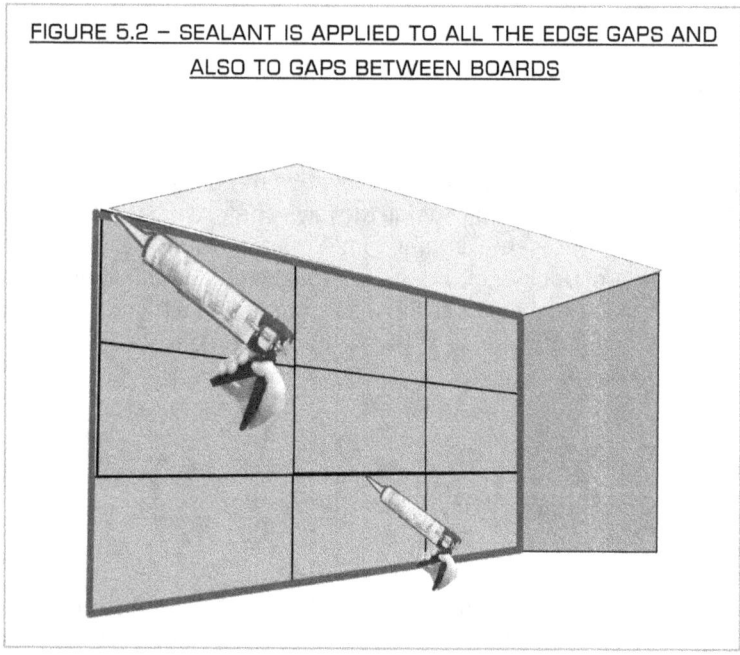

FIGURE 5.2 – SEALANT IS APPLIED TO ALL THE EDGE GAPS AND
ALSO TO GAPS BETWEEN BOARDS

The temptation to use a cheap caulking material used for filling indoor cracks, or even bathroom sealant, should be avoided. These can work acceptably for a while, but there is a risk that as they dry, they pull away from the wall edges and the seal is broken. They may also transmit more vibration as they harden,

38

Gaps between board joins should be sealed before the wall is plaster finished

Double Wall Boards

There is benefit in using double wall boards to add weight to the new wall to achieve as much noise reduction as possible. It would be best, of course, if double boards could be added on both sides of the cavity.

In screwing two boards together on the frame, they should overlap so that joins between boards do not coincide with those underneath. This minimizes the risk of an easy path for noise to leak through.

It is still good practice to seal the gaps in joins for the boards underneath before applying the second layer of boards. The gaps between joins in the top boards should be sealed before the wall is plaster finished.

Viscoelastic Adhesive

Some people advise using a viscoelastic adhesive to sandwich the boards together.

A flexible, viscoelastic adhesive applied to one surface of a board before it is sandwiched together with the other offers improved noise reduction by 'viscoelastic damping.' Specialized walls that use this kind of sandwich can achieve high STC values, but they are factory manufactured under controlled conditions.

Results for applying flexible adhesive to join two layers of boards in-situ will likely depend on how much surface is coated with the adhesive and how thick it is. Published data suggests the following can be achieved with walls using a 2 x 4 in. timber frame with studs:

(1) A wall using a single thickness of 5/8 in. (16 mm) gypsum board on one side and two 5/8 in. boards sandwiched together with adhesive on the other side results in a STC of about 55,

(2) Both walls using two double 5/8 in. (16 mm) gypsum boards sandwiched together with adhesive results in a STC of about 57.

(3) One side of the wall using double 1/2 in. (13 mm) gypsum boards sandwiched with adhesive and one single 1/2 in. wall mounted on resilient channel gives a STC of about 57.

The majority of tested wall combinations had wider stud spacings of 24 in. (600 mm) separation.

This summary has been based on results published at the following web pages:

https://www.acousticalsurfaces.com/greenglue/greenglue_test_data.htm?d=18

http://www.soundproofingcompany.com/labtests/resilient-channel-wall-with-green-glue-short-circuit-study/

The test walls appear to have used 56 oz of 'Green Glue' per 4 x 8 ft. gypsum drywall sheet pair. The frame is understood to be a cavity wall based on a 2 x 4 in. timber frame with studs 24 in. apart (not 16 in.)

Viscoelastic Adhesive versus Resilient Channel

The following three conclusions can be reached about all of this:

- The adhesive appears to produce an improvement of STC by at least 10 points compared with double boards without the adhesive.

- The biggest change in STC is achieved by applying the viscoelastic sandwich of double boards to only one side of the wall, rather than to double boards on both sides.

- Resilient channel mounting offers similar results to the viscoelastic sandwich.

Therefore, a cost-effective cavity wall construction, which would likely achieve a STC above 50, would use a single gypsum wall thickness on one side and a double wall combination on the other side (the boards being 5/8 in. (16 mm) thick) that uses either:

- resilient channel mounting of double boards on the frame, or

- a viscoelastic sandwich of double boards screwed directly to the frame,

but not both, unless you are pushing for the very highest STC value that you can reach,

Any type of adhesive which sets hard, instead of remaining flexible, would not be suitable because the vibration damping effects are lost with its loss of flexibility as it dries.

It is not known how well flexible viscoelastic adhesive will continue to work over the expected lifetime of a wall.

Supply and Handling of Boards

Gypsum wall boards should be obtained from a reputable supplier that can provide the STC values for their product.

Also, the supplier will know how to stack and store the boards properly and how to deliver them to the customer without damage. The boards can develop internal cracks which detract from their performance if they are handled carelessly.

Electrical Sockets, Switch Boxes and Ducts

Electrical socket boxes or switch boxes mounted in the cavity partition wall provide air leak paths for sound to escape from the cavity. Back-to-back boxes present the worst situation, where a neighbor's noise can pass straight through the cavity unimpeded (Figure 5.3 A).

Assuming it is not possible to do anything to your neighbor's unit, move the box on your wall at least 1 ft. (30 cm) to the side (Figure 5.3 B) if the existing cable permits. Ideally, move it to the other side of a stud (Figure 5.3 C), so that the sockets do not have line of sight to each other. Ensure the neighbor's box is covered by the sound absorptive cavity fill.

Additional noise treatment is to cover the holes in the box on your side with a mastic putty, to prevent air movement in and out of the box; the entry point for electrical cable should not be overlooked. The key is to make the box airtight.

Anything will work, so long as it fills the holes, does not support combustion and all the holes are sealed to be airtight. The filler should not shrink away when dry. Specialist material can be purchased for this; it is not necessary to cover the entire box, just the holes. If you wish, layers of flexible polymer sealant can be built up to cover the holes, each layer being added when the first has cured.

The alternative is to make an airtight box out of wallboard, that fits around the existing box, glued in place; sealant is applied to the holes where cables enter.

The box should be sealed in place in the cutout of the new wall and against the wall face. Plain, indoor sealant is good enough for a small area such as this. When it is all airtight, it will not degrade the performance of your wall.

Switch boxes for lights and ceiling fans should be similarly treated, sealing them in the cutout and against the wall or ceiling surface.

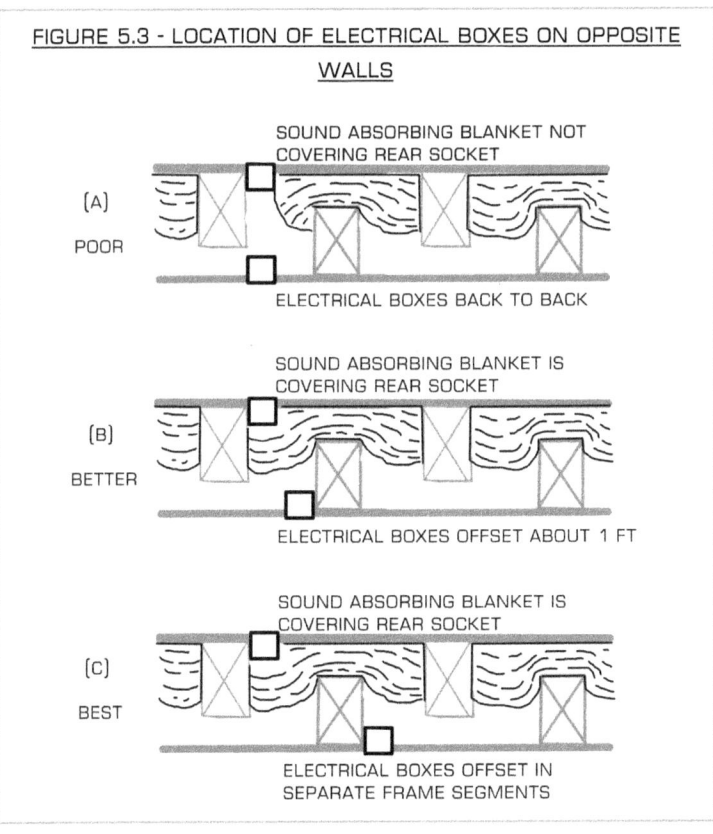

FIGURE 5.3 - LOCATION OF ELECTRICAL BOXES ON OPPOSITE WALLS

(A) POOR

SOUND ABSORBING BLANKET NOT COVERING REAR SOCKET

ELECTRICAL BOXES BACK TO BACK

(B) BETTER

SOUND ABSORBING BLANKET IS COVERING REAR SOCKET

ELECTRICAL BOXES OFFSET ABOUT 1 FT

(C) BEST

SOUND ABSORBING BLANKET IS COVERING REAR SOCKET

ELECTRICAL BOXES OFFSET IN SEPARATE FRAME SEGMENTS

43

Air Conditioning Vents

Air conditioning vents in the wall or ceiling will have ducts behind them. Ducts in the cavity are likely to be airtight or they would otherwise be leaking air into the cavity. If you find they are not airtight, then the leaks should be addressed before the wall is completed.

The mounting of a rigid, metal duct may present a vibration short circuit between two separate frames. If that is the case, attach the vent to your new frame only to reduce this possibility and prevent it from touching the neighbor's wall and existing frame (or floor and joists above in the case of a ceiling). Flexible ducting is less of a problem as it does not transmit vibration so easily.

The vent face should be sealed into the cutout of your new ceiling surface using conventional sealant, as for the electrical boxes.

6

Ceilings

Everything that has been said about cavity walls also applies to ceilings. Noise of loud music and people passes down through the floor and through your ceiling. Additionally, footfalls from the room above are a frequent cause of complaints.

The satisfactory prevention of footfall impact sounds from a neighbor's room above would involve modifying their existing floor (unless you buy them a thick carpet), which is not usually practical. The noise can be reduced, however, by improving the ceiling cavity.

> *"The overall result for a ceiling may be different from a cavity wall because it depends on the quality of flooring in the room above."*

The principles are the same as for the upgraded cavity wall, already described.

It will be necessary to take down the existing ceiling, to expose the floor above, to seal visible cracks and holes, to replace or improve the cavity insulation with 75% of fill and then to mount a new ceiling on an isolated frame or on resilient channel, or even to use both.

The overall result for a ceiling may be different from a cavity wall (which uses gypsum boards on both sides) because it depends on the quality of flooring in the room above. If you can see small gaps in the floor boards above, you should seal them if you can, as previously described in Chapter 5.

To save space, a new 2 x 4 in. (50 x 100 mm) ceiling frame should support studs which are staggered with the floor joists. Neither frame nor studs should touch the joists, so the new frame will have to be supported on the walls.

A cavity ceiling such as this has the potential to provide a STC value above 45, especially if a double layer of 5/8 in. gypsum board is used for the ceiling but, as already stated, the real-world result is at the mercy of the floor condition above and the nature of the noisy floor impacts and type of sounds involved.

Resilient Channel

The use of resilient channel, with single, or even double thickness, gypsum boards is an alternative option to building a separate frame. It can also be used in combination with the separate frame to push the potential STC beyond 50 points.

The channel should be laid across the studs of the frame, as shown in Figure 4.10 previously, to minimize contact area on the frame. The edges of the new ceiling and channel should not touch the adjoining walls. The small gaps of about 1/4 in. (6mm) at the perimeter should be filled with flexible sealant, as previously described. The gaps between joins of individual drywall boards should be sealed before plaster finishing.

Care should be taken to provide enough screws into the channel to support a heavier ceiling if double boards are used.

It will be necessary to address the back of the electrical junction box for the ceiling light (and overhead fan if fitted), which will otherwise transmit unwanted sound because it is not airtight. This can be treated as already described for electrical sockets in Chapter 5.

There may be an air conditioning vent in the ceiling with a duct behind it. The duct and vent opening should be treated as previously described.

Limitations of the Cavity Ceiling

The problem with footfall noise is that it generates much vibration which is difficult to reduce sufficiently. It may be possible for you to reduce the noise of people talking and even much of the music by, say 45 dBA, but the footfalls and other low frequency sounds may only be reduced by 10 to 20 dBA, even after careful attention to detail. Unwanted vibration from these will still enter your room by conduction through the existing walls.

Results will vary depending upon the kind of sounds produced and you might achieve better results than this. Remember that even 'only' 20 dBA reduction represents a decrease to a quarter of its previous loudness.

However, complete 'sound proofing' is not possible by treating the receiving room alone.

7

Windows

Glass is a dense material and is good at stopping noise according to the mass law: but published test data suggests that doubling the weight of glass improves the noise reduction by only 3 dB (this may have more to do with the frame structure that holds the glass in place).

Glass is available as single, double, laminated, or secondary glazed. The double glass, or dual glass comprises two panes which are usually a small distance apart to achieve good thermal insulation. It is possible to have secondary glazing, which is an additional glass pane creating a larger cavity, to improve noise reduction (Figure 7.1).

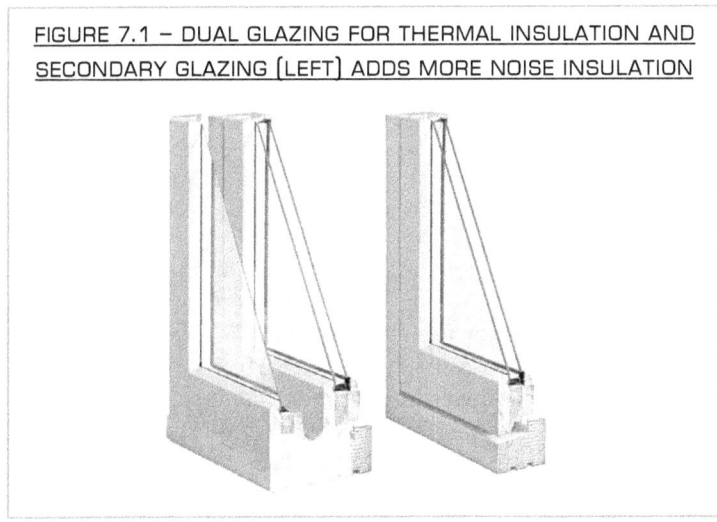

FIGURE 7.1 – DUAL GLAZING FOR THERMAL INSULATION AND SECONDARY GLAZING (LEFT) ADDS MORE NOISE INSULATION

Some vendors may quote STC values for glass panes on their own, but you need to know what the glass and the frame combined can achieve. Consider this statement in an advertisement:

> *"The glass insert has a STC rating of 38 and promises up to 95% noise reduction depending on the size of window opening, location, and type of noise."*

There are several issues here which are helpful to understand how windows reduce noise.

1. The glass has a high STC rating which gives a good impression. However, it is misleading because glass without a frame is useless and the stated STC rating may not apply to the glass and frame combination.

2. There is also the claim of 95% noise reduction but, as has already been discussed, this is a reduction of only 13 dB and is nowhere near the claimed STC rating of 38. This is scarcely better than a window of basic glass that is partially open!

---

*"95% noise reduction is only
a reduction of 13 dB."*

---

3. The vendor is claiming a high performance on the one hand and a low figure on the other, taking account of other factors which affect the result. These other factors are genuine because the type of noise really does affect the result. The STC rating (and Rw also) does not adequately describe the noise reduction performance at low frequencies, such as may come from aircraft or from road traffic with many heavy trucks and motorbikes..

4. The statement says that the noise reduction depends on the window opening size. The performance is affected mostly by the frame qualities and the seals.

The lessons to learn from this are to beware of the deceptive *"stops 95% of all sound"* claims and to ensure you are quoted STC values for glass and frame together.

In choosing windows, it is better to maintain a balance of window performance and walls, by aiming for STC or Rw values that do not greatly exceed the STC value of your existing walls. Since a residential wall is likely to have a STC of 35 to 45, that would suggest the kind of performance to seek for your windows if your budget will permit.

Gaps that leak air

Real-world noise performance depends on how well the window frame has been fitted into the wall of your home.

> *"Real-world noise performance depends on how well the window frame has been fitted into the wall."*

If the frame has many air gaps around it, it will let more sound through, and you will not benefit from the manufacturer's claimed noise reduction for a perfect fit.

This is frequently a problem with existing windows in your home. They might be good thermal insulators (because they are double glass) but the attachment of the frame to the house wall may leave air gaps which counter the noise performance of the window and frame. If there is a problem, you will be able to hear sound leaking through when you place your ear close to the edge of the frame.

It is impossible to address these gaps without removing the existing wall surfaces around the frames, either inside or out. Understandably, not everyone is willing to do that. If you do, fill the gaps with a good quality flexible polymer seal, as used for external applications (Figure 7.2).

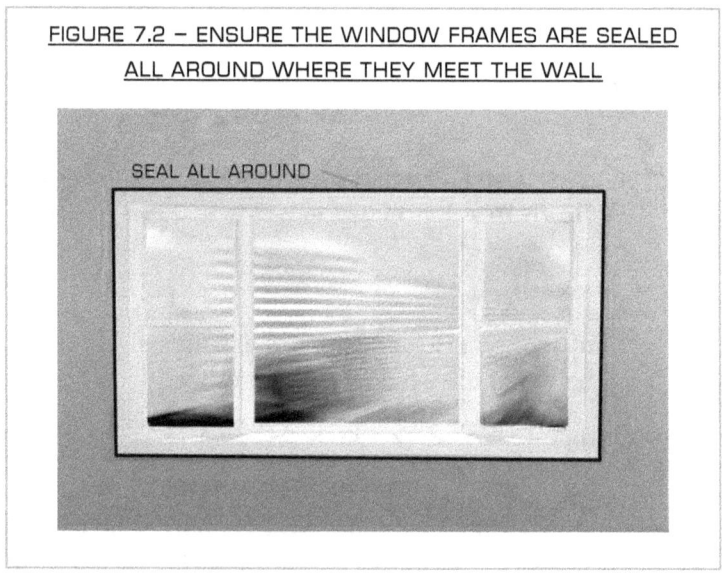

FIGURE 7.2 – ENSURE THE WINDOW FRAMES ARE SEALED ALL AROUND WHERE THEY MEET THE WALL

SEAL ALL AROUND

Existing sliding glass doors can have poor noise performance because the weather seals are damaged or degraded. Replacing or upgrading these seals (where possible) can improve the noise integrity of the whole door when it is closed.

The same applies to casement windows: check how well they seal when closed. Improvement to the sealing with after-market material can restore their original performance.

The gap between panes of glass

Many standard windows purchased from hardware stores appear to use windows of dual glass each 3/32 to 1/8 in. thick (3mm or less). If STC values are provided at all, they would likely be around the mid 30's when well fitted in place and assuming the seals on closures are good. In the context of noise reduction, these will be better than an old leaky frame with single glass.

A narrow gap between the glass panes improves thermal efficiency, but noise control requires a larger gap. A rule of thumb is that each doubling of the gap between the panes increases the STC of the window by 3 points. However, this works against the thermal insulation properties of the window.

Secondary glazing is used to boost the noise performance for low frequency noise and to maintain the thermal insulation of the dual glass: an extra glass window is added on the inside of the frame.

A cavity space of at least 2 in. (50 mm) is required to achieve significant reduction of road traffic and aircraft noise. Better results are obtained if it is 4 in. (100 mm). or even more.

A sliding window of 4mm glass is feasible for after-market improvement of existing windows, although the air gap will depend on the space available on the internal sill. Attention should be paid to the seals on the sliding frames to minimize air leakages when the windows are closed. It is possible to achieve windows which produce STC values in the high 30's with this kind of treatment.

Double glass with different thickness of panes work better than two panes of equal thickness, but the cost is higher. Two different glass thicknesses help to spread out the noise reduction across all the frequency spectrum, without a pronounced degraded performance at one frequency in the most sensitive range of human hearing.

Laminated Glass

Laminated glass is sometimes used to improve thermal efficiency and reduce unwanted heat or light from the sun. The noise benefit comes mostly from the added weight of two panes of glass joined together, rather than from the lamination itself. Sometimes, lamination joins two glass panes of different thickness.

The lamination process itself only improves the noise reduction of the window at higher frequencies by about 2 dB, so lamination should not be specified for noise improvement alone unless it is used to gain the extra weight and different thicknesses of glass.

Summary

In summary, the trend for improved noise reduction of windows is found with:

- thicker glass (because the weight is increased). Doubling the weight of glass increases window and frame STC by 3 points;

- increasing the cavity between the glass. A doubling of cavity width increases the STC by about 3 points (but this degrades the thermal performance of the window);

- differing glass thickness for the two panes (if possible);

- a good, tight fit of opening windows to the frame when closed (the weather seal is important);

- the tight seal of sliding windows or doors in their channel and a tight seal of the frames when closed;

- quality fit of the frame into the surrounding walls of the house, with sealant to fill air gaps.

Some typical values of STC performances are shown in Table 7.1, overleaf. (Rw values are similar.) These provide only a broad guide because reported results for individual window designs vary enormously among vendors for similar glass combinations.

TABLE 7.1 – TYPICAL STC VALUES OF GLASS COMBINATIONS

Glass description	STC
1/8" (3mm)	29
1/4" (6mm)	32
1/8" (3mm) + 1/8" (3mm) laminated	35
1/4" (6mm) + 1/4" (6mm) laminated	38
1/8" (3mm) + 1/4" (6mm) air gap + 1/8" (3mm)	28
1/8" (3mm) + 1/2" (12mm) air gap + 1/8" (3mm)	31
1/8" (3mm) + 1" (25mm) air gap + 1/8" (3mm)	35
1/8" (3mm) + 1/2" (12mm) air gap + 1/4" (6mm)	37
1/8" (3mm) + 1/8" (3mm) laminated + ½" (12mm) air gap + 0.2" (5mm)	42
1/2" (12mm) + 1/4" (6mm) laminated + 6" (150mm) air gap + 1/4" (6mm) + 1/4" (6mm) laminated	50

If you are an architect reading this book, be aware that noise performance of windows is a technology of its own and I have only been able to cover the most significant items that trend towards improved noise reduction.

It is important when specifying new windows to obtain precise vendor test data for a window and its frame combination to be sure that it will perform adequately in the anticipated noise environment. It is also important to specify how the frame is attached to the building and to supervise its installation.

8

Doors

Doors present a similar problem to windows, because they must open and close and they leak sound where they meet the door frame. The door is often the last item to be considered when reducing noise from the external environment, after walls and windows have been dealt with. It truly represents the weakest link in an external wall because it is difficult to achieve a similar level of STC as the wall without very great expense. Fortunately, the door has a small surface area compared with the wall and so its weakness need not be overwhelming.

For a door to reduce unwanted noise successfully, the seals around the head, jamb and floor sill must be airtight for its service life. A proprietary storm door with tight fitting seals can do a good job at reducing the transmission of external noise to inside the home. Unfortunately, inexpensive storm doors may appear tough but are usually hollow. A door which contains a cavity will allow much external sound to pass straight through and is unlikely to achieve a STC higher than about 25. More expensive doors, sometimes using solid hardwood, offer improved noise reduction because of their weight and lack of internal cavities, in addition to quality sealing.

Internal doors are often light in weight and do not fit tightly against their frames when closed; significant gaps at the floor allow much sound to pass between rooms. If room-to-room sound transmission is a problem it will be insufficient only to replace the door with a heavier, more solid version: the frame sealing should also be addressed in the same manner as for a storm door on the outside wall.

The after-market has numerous items for improving the sealing of doors against drafts and these are also the most cost-effective way of sealing them against noise. You have to ensure the door is sealing against the top and side frames, including the hinged side of the door, and to 'draft proof' the bottom of the door.

Some guideline values of STC for doors are given in Table 8.1, below, but the quality of the seal of the door against its frame when it is closed is often the controlling factor for the total STC achieved.

TABLE 8.1 - STC VALUES OF SOME DOOR DESIGNS

Material	Core	Glass	STC
Wood or MDF	hollow	none	20
Steel	foam	none	23
Wood	particle board	none	29
Fiberglass	foam	1" (50 mm)	25
Fiberglass	particle board	1" + ¼" (50 mm + 6 mm) laminated	29
Fiberglass	particle board	1" + ¼" (50 mm + 6 mm) laminated	30
MDF + Wood Fire Stop 45 mins	solid	none	40

Internal fire-stop doors seal the door in its frame and at the sill underneath.

9

External Air Conditioning

The external air conditioning unit can pose a problem from one neighbor to another, and for you or your neighbors who wish to be outside enjoying the yard or patio. It can also cause excessive noise to enter your home.

There are three sources of noise in the external unit:

- compressor,
- fan,
- body.

The compressor vibrates and generates a lot of noise at low and middle range frequencies and the fan rotates at high speed, generating high frequency air-movement noise. The body of the unit picks up much of this noise as vibration and retransmits it as sound into the open air.

Manufacturers of new units have managed to address much of the noise emitted by older models: the compressor is better enclosed in noise resisting material and is mounted on vibration damping mounts, fans are of improved design and some units incorporate a variable speed so they do not have to run at full speed all the time.

However, even with new units, some unwanted noise remains. There are all kinds of suggestions out there about how to build fences or hang 'blankets' on fences to prevent the sound. Some of the ideas are a waste of money because they are placed too far from the unit.

It is a principle of noise control that you should:

(1) first address the noise at source: make sure the unit is well maintained and the body or fan guards are not vibrating unnecessarily (it almost goes without saying, but parts do work loose over time and you don't notice the steady increase in noise that occurs);

(2) deflect the noise away from the annoyance with a barrier wall (i.e. a fence) and place it as close to the unit as possible, but allow adequate space for ventilation.

You cannot stop the noise completely, but you can redirect it to reduce noise in the appropriate direction. For example, if noise from the unit is projecting towards an upstairs window, you will need to fit a hood over the unit to direct the fan noise elsewhere, but not so close that the free air flow is affected. If the noise is a problem on one side, then a barrier wall is required on that side of the unit.

The barrier can be as simple as a wooden fence, but it should be constructed of material that does not have holes (which otherwise let too much sound through) and is firmly supported against anticipated wind. It will only work if it is placed close to the unit, yet not so close as to affect air flow.

Figure 9.1 shows an example of where the barrier should be positioned to reduce a noise problem on the neighbor's side, together with the scale required.

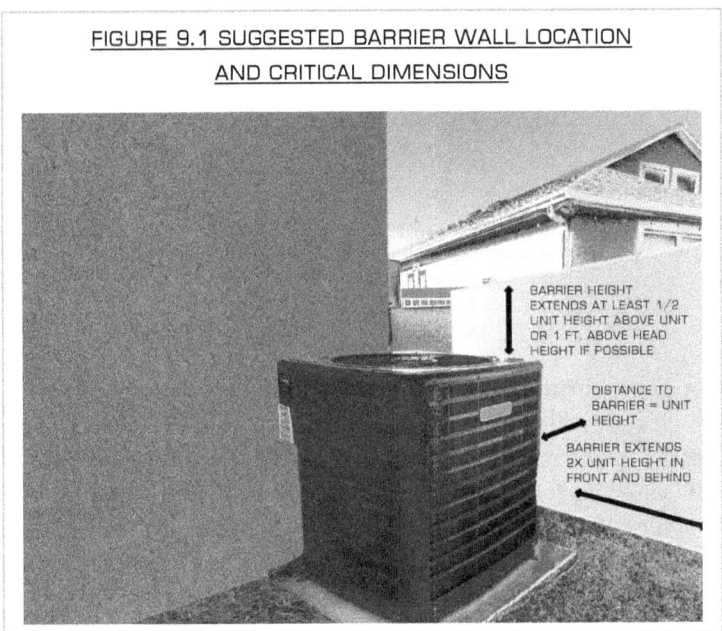

FIGURE 9.1 SUGGESTED BARRIER WALL LOCATION
AND CRITICAL DIMENSIONS

BARRIER HEIGHT
EXTENDS AT LEAST 1/2
UNIT HEIGHT ABOVE UNIT
OR 1 FT. ABOVE HEAD
HEIGHT IF POSSIBLE

DISTANCE TO
BARRIER = UNIT
HEIGHT

BARRIER EXTENDS
2X UNIT HEIGHT IN
FRONT AND BEHIND

It will not be perfect because noise will still escape above the barrier and around the sides.

If the person being affected just has line of sight above the barrier to the ac unit (or at one side edge of the barrier), the noise reduction will be 5 dB. A further reduction of 3 dB can be expected for each foot (about 1/3 meter) extension of the barrier beyond line of sight, up to a maximum of about 12 dB.

In the example shown in Figure 9.1, it would be possible to reduce the length of the barrier behind the unit by constructing an L-shape to the wall of the house (Figure 9.2).

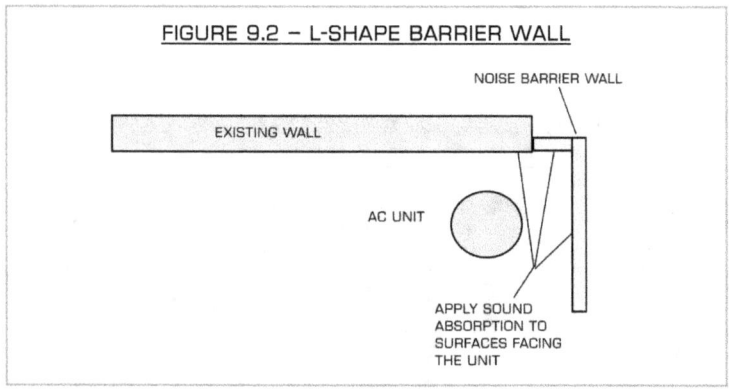

FIGURE 9.2 – L-SHAPE BARRIER WALL

NOISE BARRIER WALL

EXISTING WALL

AC UNIT

APPLY SOUND
ABSORPTION TO
SURFACES FACING
THE UNIT

You can improve the barrier performance by reducing reflections. Noise is already reflected from the wall behind the unit and it will now be reflected from the barrier, possibly towards your other neighbor or towards your back door! There is no benefit in attaching sound absorptive material to your neighbor's fence because it is likely to be too far away; it should be applied to the surface of the barrier wall facing the unit.

Mineral fiber (e.g. Rockwool) or glass fiber insulation absorbs sound and is used indoors where it is dry, but outside can absorb moisture and get very wet in the rain, becoming a haven for moss, fungi and bacteria. If it is covered with plastic sheet to keep the rain off, its sound absorptive properties are greatly reduced.

Nevertheless, it is possible to use it outside if it is mounted on a firm backing, such as metal or plywood and covered with a perforated or mesh grid to prevent physical damage, but it must have free drainage for excess moisture to escape. Mineral or glass fiber blanket that is at least 2" (12mm) thick should be used.

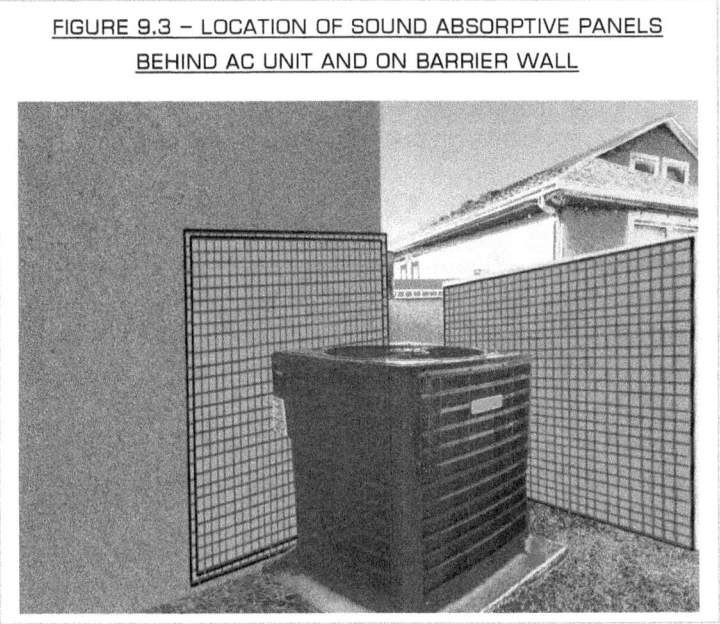

FIGURE 9.3 – LOCATION OF SOUND ABSORPTIVE PANELS BEHIND AC UNIT AND ON BARRIER WALL

Figure 9.3 shows where sound absorption should be applied (hatched lines).

Practical longevity of sound absorptive material outdoors depends on weather conditions in your area. Ideally, it should be able to dry quickly in the sunshine after getting soaked by rain.

If foam is used, instead of glass fiber blanket or mineral fiber, it should be open foam, rather than closed foam. (You can tell if the foam is 'open' because, if you hold it to your lips, you can blow through it.)

Vendors of noise control products can supply sound absorptive foam which has been covered with an extremely thin melamine finish. This is thin enough to keep the foam moisture-resistant and yet able to absorb sound.

If it is necessary to prevent noise escaping on all sides, construct a three-sided barrier, making use of the existing house wall, as shown in Figure 9.4. Noise will be directed upwards with this approach, which is not always desirable. A small gap (say 6 inches (150 mm) beneath the barrier all around should be used to improve air flow.

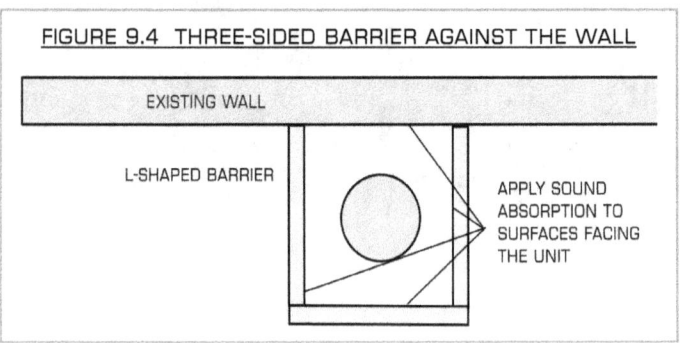

FIGURE 9.4 THREE-SIDED BARRIER AGAINST THE WALL

EXISTING WALL

L-SHAPED BARRIER

APPLY SOUND ABSORPTION TO SURFACES FACING THE UNIT

Advice that appears frequently on the internet, but which will not work, is to:

- *"apply sound absorption to the neighbor's fence"* –the neighbor's fence is usually too far away;
- *"plant a line of shrubs or trees in front of the unit"* – tall vegetation would have to be planted tens of feet deep to achieve a noticeable difference.

10

Aircraft Noise

Protecting your home against aircraft noise takes the requirements for noise reduction to a whole new level (forgive the pun). Aircraft noise differs from other intrusive noise in its intensity, which occurs in short bursts. More extreme precautions are necessary to reduce this noise and it is an expensive process. The entire house has to be made more airtight: roof, walls, windows, doors and even ventilators require attention.

Internet web sites carry some useful publications about the level of detail required and a list of references to some of these is provided in Chapter 12. You should use these as your guidelines for reducing aircraft noise, because they would otherwise require another book devoted to this single subject.

Windows

Immediate benefit can be achieved in an existing house by improving the windows. If you can add storm-sealed windows to the outside of existing old, leaky windows, you can boost their STC performance into the 30's, although it depends on how well the frames fit with the surrounding walls. However, if higher insulation is required, it may be cheaper in the longer term gradually to replace old windows with new units that offer combined advantages of improved thermal and noise insulation.

Roof and Eaves

The noise that enters a room through the roof and ceiling can be treated by improving the ceiling, as already described. However, even ceiling treatment may not be enough on its own, which then requires attention to how noise is entering the roof space.

Noise enters directly through the roof structure, as it does through any wall, except that the roof is usually constructed of lighter materials and therefore transmits more noise. There are likely to be many air spaces where the roof meets the walls at the eaves. The eaves are the priority for noise control because these air spaces allow the noise to leak directly into the loft space and are large in total area.

It is not simply a case of filling the gaps with thermal insulation (much of the low frequency sound passes straight through without reduction), but instead of building a sound resisting structure, not unlike a cavity wall. Heavy, mass-loaded vinyl is helpful in tight spaces such as these, because it is flexible and can be bent at the edges to seal where the roof meets the wall. It can be built up in layers, if necessary.

The awkward position of eaves usually requires the assistance of professional builders to complete the task satisfactorily.

Noise penetration through the roof itself can be reduced by adding particle board to the roof joists, with sound absorption inside the space, just like a cavity wall. However, precautions concerning weight loading of the structure and winter condensation issues in cold climates are likely to require professional help before the roof is treated in this way.

Ceiling

If the ceiling joists are still exposed in an attic or loft, the noise penetrating the ceiling of the room below can be reduced by adding a floor on top of the joists (usually of particle board strong enough to walk on), with sound absorption inside the cavity. Effectively, a cavity ceiling is created for the room beneath. The usual precaution is required of adding flexible sealant to fill potential air leaks between individual boards as they are placed in position.

Ventilators

Open ventilators in the walls of the home allow noise to enter easily. They can be provided with a baffle box of wood or metal that covers the vent but allows air to pass through a slot underneath (figure 10.1). The inner surfaces are lined with at least 2 in. (50 mm) of sound absorption, which is either open foam or mineral or glass fiber blanket. Naturally, enough space should be provided for ventilation air to pass: an air gap of about 4 in. (100 mm), allowing also for the absorption thickness, would be suitable.

More complex versions for extra noise reduction are described in the reference literature.

It is sometimes necessary to fit special silencers for vents that project vertically through the roof.

FIGURE 10.1 – A BAFFLE BOX ALLOWS AIR TO VENT
FROM ONE DIRECTION BUT NOISE TO BE REDUCED

SLOPING ROOF, SIDES FITTED;
INNER FACES OF BOX COVERED
WITH SOUND ABSORPTION

AIR VENTS
THROUGH
SLOT
UNDERNEATH

Health

When the home has become sealed against noise, there is the issue of its heating and ventilating. Central air conditioning is common for houses located near airports because it minimizes the need to open windows and inadvertently let the sound in.

The health implications of sealing the home will require advice from a heating and ventilating professional. The build-up of carbon monoxide from gas boilers or heaters and even from radon gas in some locations are serious issues to consider.

11

SUMMARY OF NOISE CONTROLS

Table 11.1 Walls and Ceiling

The more items that you can check off (√) for your proposed improvements, the better.

Improvement	Noise Reduction	Notes	√
Increasing total weight of wall improves noise reduction.	Double the weight of a wall improves noise reduction by 5-6 dB. A 50% increase in weight improves it by 3 dB.	Can be a challenge to find the space and to support a heavier wall or ceiling.	
Maximize the cavity between walls	A doubling of cavity space increases the noise reduction by at least 3 dB.	Requires more space.	
Fill the cavity to a depth of 75% with sound absorption.	Glass fiber or mineral fiber (e.g. Rockwool) improves noise reduction by 3-5 dB	No significant improvement if the cavity is filled 100%, except for thermal insulation improvement.	

Improvement	Noise Reduction	Notes	√
Wider frame spacing outperforms 'standard' frame spacing.	24" stud separation instead of 16" improves noise reduction by 3-5 dB.	A useful benefit in combination with other improvements.	
Mount the internal wall or ceiling on a resilient channel.	Reduces vibration transmission and contact area; improves noise reduction by 10 dB.	In general, the lighter the gauge of metal for the channel, the better it performs. 25 gauge metal is commonly used.	
Two wall boards can be 'sandwiched' with flexible viscoelastic adhesive	Reduction of about 7 to 10 dB compared with two boards fixed together without sandwich	Depends on thickness and total cover of adhesive. Diminishing returns if combined with other noise control features, such as resilient channel mount.	
		For best cost efficiency, use either viscoelastic sandwich or resilient mount for double boards, not both.	
Wall boards should not touch adjoining walls, floor or ceiling.	Improves noise reduction by 6 dB; reduces vibration transmission.	Allow a ¼" (6mm) gap; fill edges with flexible sealant.	

Table 11.2   Walls and Ceiling: Good Practice

Improvement	Notes	Additional Information
Seal board joints before plastering	May improve noise reduction by 3 dB.	Good practice
If doubled wall boards are used:	(1) wallboards should be staggered relative to boards underneath; (2) boards of different thickness are preferred, but don't do this if it means reducing the intended wall thickness (weight).	(1) staggering helps to seal joins underneath; (2) differing thickness improves the 'spread' of sound frequency performance and overcomes limitations of each board alone.
Address electrical boxes	Move opposing boxes if possible. Make airtight. Fill holes in the back, seal the front where it meets the wall.	Overcomes a weakness that detracts from the wall or ceiling treatment. Make the boxes airtight is the key to this.
Address air vent ducts in cavities	Ensure rigid duct is not connecting new and old frames (if new frame fitted). Seal the vent where it meets the wall or ceiling.	Prevents vibration short circuit and leakage of sound into room.

Table 11.3   Typical STC Values of Cavity Walls

Your results of noise reduction will likely vary from these figures depending on the nature of the sound, the materials used and how much attention has been paid to details, as well as how much vibration enters the frame from surrounding walls, floor or ceiling.

Description	Sketch	STC
5/8" (16mm) gypsum board has a weight of 2 lb/ft² (10 kg/m²)		
Basic cavity wall — 2"x4" (50x100mm) timber studs, 16" (400mm) centers, 75% glass fiber fill, 5/8" (16mm) gypsum board		35
Wider stud spacing — 2"x4" (50x100mm) timber studs, 24" (600mm) centers, 75% glass fiber fill, 5/8" (16mm) gypsum board		40
Staggered studs — 2"x4" (50x100mm) timber studs, 16" (400mm) centers staggered on a 2"x6" (50x150mm) frame, 75% glass fiber fill, 5/8" (16mm) gypsum board		45
Separate frames, staggered studs, double wall thickness — 2"x4" (50x100mm) timber studs on separate frames, 16" (400mm) centers staggered, 75% glass fiber fill, double 5/8" (16mm) gypsum boards on both walls		58

Description		Sketch	STC
Single wall board on resilient channel	2"x4" (50x100mm) timber studs, 16" (400mm) centers, 75% glass fiber fill, 5/8" (16mm) gypsum board on each wall, (one resilient mounted).		48
Double wall board on resilient channel	2"x4" (50x100mm) timber studs, 16" (400mm) centers, double 5/8" (16mm) gypsum board on inside wall (resilient mounted), 75% glass fiber fill.		55
Steel stud basic wall	3 5/8" (90mm) steel studs, 16" (400mm) centers, 75% glass fiber fill, 5/8" (16mm) gypsum board		40
Steel stud resilient channel	3 5/8" (90mm) steel studs, 16" (400mm) centers, 75% glass fiber fill, 5/8" (16mm) gypsum board; inside wall on resilient channel		48
Steel stud deeper cavity, resilient channel	6" (150mm) steel studs, 16" (400mm) centers, 75% glass fiber fill, 5/8" (16mm) gypsum board, inside wall on resilient channel		53

73

Table 11.4   Improving Windows Noise Reduction

Improvement	Noise Reduction	Notes	√
Increasing total weight of glass improves noise reduction.	Double the weight of glass improves noise reduction by 3 dB.	Sometimes achieved with secondary glazing, or with lamination.	
Maximize the cavity distance.	A doubling of cavity space increases the noise reduction by 3 dB.	Requires more space. Less effective for thermal insulation.	
Different glass thicknesses improves noise reduction.		Usually achieved with lamination, or secondary glazing.	
Laminated glass	Can improve higher frequency reduction by about 2 dB	Usually applied for reasons other than noise	
Add secondary glazing to existing window.	Doubling cavity space and doubling glass weight together increases noise reduction by 6 dB.	Requires more space. Sealing of secondary glazing should be airtight when closed.	
Ensure window seals tightly when closed.		Poor, or damaged seals leak noise.	
Ensure the frame fits well with house walls.	Vital to achieve performance stated by vendor	Use sealant where the frame meets the wall to fill in unwanted air gaps.	

Table 11.5   Typical Range of STC Values for Windows

There is a very large variation in reported STC values for windows. Results differ because of different frame designs, attention to sealing and different glass combinations. In the end, your results of noise reduction will likely differ from the figures indicated by vendors, depending on the nature of the noise and how well the frames fit into the surrounding walls.

Window Type	Frame Sealing	STC
Old 'leaky' windows	Poor seals; windows with better STC have improved sealing with frame and with surrounding walls	20 - 25
New single glazed frame	Good sealing with frame	25 - 35
Double glass in frame	Storm sealing with frame	35 +
Triple glass in frame	Storm sealing with frame	40 +

Table 11.6   Typical Range of STC values for Doors

The results will vary between vendors for different styles and construction of doors. The STC value is highly dependent on how well the door seals into its frame when closed.

Door Type	Frame Sealing	STC
Lightweight with empty cavity or foam core	Poor seals; doors with better STC have improved sealing with frame	20 - 25
Heavier door with solid or particle board fill	Good sealing with frame	30 +
Fire stop door	Very good sealing with frame	40 +

Table 11.7   External Air Conditioning Unit

Feature	Comment
Barrier wall (or even a hood over the top) can be made of any rigid material without holes, that can be supported against wind.	A timber fence is ideal
Place barrier as close to ac unit as is feasible for air circulation	In the example, the distance is equal to the height of the ac unit.
Sound absorption, at least 2 in. (50 mm) thick, should be applied to the surfaces facing the ac unit	
The barrier should be as high as practical	Just line of sight over the barrier gives 5 dB reduction, improving by 3 dB for each foot (30 cm) above the listener
The barrier should extend as far as practical beyond the ac unit, front and back.	As above, the reduction improves by 3 dB for each foot (30 cm) beyond the listener's line of sight.
A noise barrier wall (or fence) is unlikely to achieve more than about 12 dB reduction.	

Table 11.8   Aircraft Noise

Summary Notes
Refer to the references in Chapter 12 for the detailed attention that may be necessary
Roof, walls, windows, doors and even ventilators require attention.
The eaves are the priority for treatment in the roof because these are 'leaky.'
Immediate benefit can be achieved in an existing house by improving the windows if they are old and leaky, but this may not be enough on its own to reduce intense and frequent aircraft noise.

12

Helpful References

Walls and Windows

"The Sound Book," National Gypsum, USA
https://www.nationalgypsum.com/file/THESOUNDBOOK.pdf

"Drywall Manual," Siniat, UK
https://www.siniat.co.uk/en/knowledge-centre/drywall-manual

"Sound Transmission Class Guidance," U.S. Dept. of Housing and Urban Development (HUD)
https://www.hud.gov/sites/documents/DOC_16419.PDF

"STC Chart," John Sayers, Australia
http://johnlsayers.com/Recmanual/Pages/STC%20Chart.htm

"Sound Abatement Windows Can Help Make a House Hushed and Healthy," Eco Building: Pulse, American Institute of Architects
http://www.ecobuildingpulse.com/products/sound-abatement-windows-can-help-make-a-house-hushed-and-healthy_o

Aircraft Noise

"Tips for Insulating Your Home Against Aircraft Noise,"
Metropolitan Airports Commission, USA
https://www.macnoise.com/sites/macnoise.com/files/pdf/tips.pdf

"Sound Insulating Your Home," City of Chicago, USA
http://dot.ca.gov/hq/planning/aeronaut/documents/soundinsul.pdf

"Reducing Aircraft Noise in Existing Homes," Perth Airport,
Australia
http://www.perthairport.com.au/Files/Reducing_Aircraft_Noise_in_Existing_Homes.pdf

"Reducing Impact of Aircraft Noise at Home," Air Services,
Australia
http://www.airservicesaustralia.com/wp-content/uploads/15-151FAC_Reducing-impact-of-aircraft-noise-at-home_WEB.pdf

"How to Soundproof a Roof for Traffic and Aircraft Noise,"
soundproofingexpert
https://youtu.be/ea9dCz42hSo

Index

www.ingramcontent.com/pod-product-compliance
Lightning Source LLC
Chambersburg PA
CBHW071220220526
45468CB00002B/690